SAILBOARDS
custom-made

Hans Fichtner and Michael Garff

SAILBOARDS

custom-made

Second Edition

Building and Designing
your own Sailboard

Stanford Maritime
London

Stanford Maritime Limited
Member Company of the George Philip Group
12–14 Long Acre   London WC2E 9LP
Editor Phoebe Mason

First published in Great Britain 1982
English edition © Buchheim Editions SA
and Stanford Maritime Limited 1982

Second edition © Stanford Maritime 1986

Set in Monophoto 11/12 Univers 689 by
Tameside Filmsetting Ltd, Ashton-under-Lyne, Lancs.
Printed in Great Britain by
Butler & Tanner Ltd, Frome, Somerset
Cover photo by Cliff Webb/Pickthall Picture Library

British Library Cataloguing in Publication Data
Fichtner, Hans
    Sailboards Custom-made.
    1. Windsurfers—Design and construction
    I. Title   II. Garff, Michael
    III. Surfboard custom-made. *English*
    623.8'22   VM351

ISBN 0-540-07317-2

Contents

Acknowledgement

For their friendly assistance in the
preparation of this Second Edition,
the Publishers would like to thank:

Shore Boardsailing
Unit 12, Wittering Walk, East Wittering,
Chichester, Sussex

Structural Polymer Systems Ltd
Love Lane, Cowes, Isle of Wight

Materials

Core blanks

Two different plastics, polystyrene and polyurethane, are used for the foam cores of sailboards. Both are manufactured in large blocks or slabs for commercial purposes and may have to be cut to size; sometimes a block can be shared between two or more boards. Kit suppliers offer various suitable blanks also.

Polystyrene foam

Expanded polystyrene (EPS) is a general term for the type of foam made by heating and expanding small polystyrene beads or pellets (Styropor) which fuse together into foam of the required density. A wide range of densities can be produced, and the end product is widely used for packaging, building insulation and buoyancy. The white beads can be coloured before expansion, usually giving a slightly mottled result, or the surface of the foam can be coloured. The ultraviolet in sunlight eventually causes yellowing.

An average density of 20–30 kg/cu m is most suitable for sailboards; this is far harder and stronger than the grades seen in packaging and may be known as 'extra high density' (EHD) foam. Although there are densities of up to 60 or even 100 kg/cu m, 20–30 is quite adequate to give the resistance to compression and the strength needed for sailboards. At this density, and if the surface is covered with a properly laminated and finished fibreglass skin, the degree of water absorption is not significant, though it is greater than for extruded PS.

Expanded polystyrene can be made in moulds shaped as blanks for sailboards or surfboards, which gives a firmer surface and saves on the cutting-down and smoothing work, though pre-determining the shape and maximum dimensions.

Extruded polystyrene is a stronger and much more commonly used type of polystyrene foam, made by a different process and sold under the names Styrodur (BASF) or Styrofoam (Dow), among others. It again is manufactured in large blocks and sheet, largely for the building industry, and is a closed-cell foam available in different densities. It is not marketed in moulded sailboard core shapes, however, and one would have to cut it from a standard sized sheet or block. The same density of 20–30 kg/cu m is desirable; the cost may be slightly more than EPS made from beads. However, the foam is more resistant to water penetration than EPS, which although also closed-cell may take up some water between the beads, an eventual possibility if the board's skin should be punctured or cracked. Colour can be added in manufacture or afterwards.

Both types of EPS are very fragile because of their porous structure and lack of stiffness. However, they are easily worked with normal hand or power woodworking tools and can be cut by sawing or with a hot-wire saw which one can make (described below). Especially with foam which has the bead structure, it is important that the tools for cutting and smoothing are not too coarse or the beads could be torn out of the surface, roughening it; hand saws and 120 grain sandpaper can be used. It is well worth practising on scrap pieces first to see which tools and sandpapers work best, and to perfect your technique.

Epoxy resin must be used when laying-up fibreglass over polystyrene foam as the styrene in polyester resins attacks and dissolves it. Filler pastes for repairs must not contain polyester, but two-part PU foam can be used safely.

Polyurethane foam

Rigid polyurethane foam is also used for sailboards, and the fibreglass skin can be laminated with either polyester or epoxy resin. However, fittings should be put in with epoxy, for better bonding.

PU foam is also primarily manufactured for purposes such as building insulation or sandwich construction, and you may have to cut down a standard sized block. It is foamed from two components, designed to produce a range of properties and densities, and is a closed-cell foam with good resistance to water penetration, stiffness and firmness. It is easy to work with, but *must be cut by sawing* (with a fine-tooth hand saw) and not by hot-wiring which produces very poisonous fumes.

The more usual PU (and PS) foams may be coloured in manufacture, though with clever use of graphics this may not matter.

An alternative, though more expensive, is the PU foams sold under the tradenames Clark (from the USA) and Bennett or Burford (from Australia). They are white, and have uniform density and freedom from voids, which enhances strength and resistance to water penetration. Blanks may be available pre-stringered and/or partly shaped. They are imported in various sizes and handled by kit makers such as Shore Boardsailing.

Making your choice

If you are unable to obtain or afford the specialized foams, then Styrofoam or Styrodur extruded polystyrene is the next best. The extra strength contributed by epoxy resin laminating will give you a very durable result. The least desirable option is the weaker expanded-bead polystyrene, but boards have been made successfully with it using strong epoxy layups, and it may be all you can find. For use where the board is not going to be subjected to much stress, e.g. for a beginner's or child's board, the extra high density grade can be quite adequate.

Laminating materials

After stringering, shaping and applying the graphics, the board gets its 'skin', laminated with resin and glass fibre or other reinforcement. Epoxy and polyester are the two resins used, though the latter can only be applied to PU cores.

Reinforcement

The fibreglass or GRP on the outside of a board is a composite material in which the various properties of both the resin and its reinforcement interact.

Fibre reinforcement will increase the impact resistance and strength, reduce shrinkage and stabilize the heat resistance of the resin. E-glass fibre in the form of fine tissue, mat or as woven cloth is used almost exclusively for the reinforcement which is so important for board construction, as it has good water and alkali resistance. The strongest and most common form is woven glass cloth with a weight of 200–240 g/sq m or 6–8 oz/yd; various widths are available and it is worth seeking one that needs least cutting to size and wastage.

Carbon fibre reinforcement is way ahead in terms of tensile strength. Properly used, it gives a board more stiffness, but is much more expensive than glassfibre. Also expensive, and difficult to work, is Kevlar reinforcement. This material used in combination with glassfibre achieves a very high level of stiffness as a result of superimposed layer laminating. It can also be used as the sole reinforcement, seen in ultra-light boards, and sometimes in rowing shells and racing dinghies. Kevlar and carbon fibre reinforcement are really only worth the price in the building of very light, highly stressed boards or boats, or when cost is not limited; they may also be forbidden by class rules.

These 'exotics' have to be used

with more regard to their particular properties than glassfibre, to justify their cost and greater difficulty in handling. Both Kevlar (aramid) and carbon fibre have high tensile strengths for their weight, and can be applied as a strip of unidirectional fibres running lengthwise on the hull (or on a daggerboard), in conjunction with a glass layup. Kevlar cloth is also used. Carbon is very vulnerable to dents and nicks, which destroy its strength, so it has to be laid underneath protective layers of glass and/or Kevlar, which both have good impact resistance. It must not be cut across by slots for footstraps or other fittings.

Kevlar has poor compression strength, however, so it is used with glass on decks. It cannot be sanded smooth, so the outer layer(s) is usually the much more easily handled glass.

Reinforcements of all types are often pre-treated to repel moisture and improve bonding with resins. It is worth checking that your supply is suited to the resin you are using. Some manufacturers, such as SP Systems or Strand Glass, supply both resins and reinforcements and can advise and offer compatible systems.

Polyester resins and laminates

Among the many laminating resins on the market, the unsaturated polyester resins are most commonly used for boats and boards. The molecules in hardened resin are not in the form of a chain but three-dimensionally crosslinked. Furthermore, the structure cannot be dissolved by heating: it is thermosetting rather than thermoplastic. This means that nothing can be changed on the board after the resin has hardened —any wrinkles, bumps and bubbles remain—so with all resins one has to work very exactly.

Polyester resins come either in straight liquid form or with additives, which include fillers to extend the bulk or a thickening or thixotropic agent such as aerosil, also colouring pigment. Various fibrous substances are added to make paste fillers, and fine glass fibres are recommended for use on sailboards, to be applied with a spatula. Colloidal silica and microballoons can be mixed into the resin as fillers to achieve a smooth surface finish.

In order to activate polyester so that it will react, something must be added to the resin. As the resin is transformed from a liquid to a gel and the hardens to a solid, heat of polymerization is given off; i.e. the process is exothermic. Should liquid leak out, normally styrenes, it has nothing to do with the polymerization and hardening of the resin. On the contrary, the setting of unsaturated polyester resins is induced by a catalyst or hardener, usually an organic peroxide such as MEKP, and an accelerator which speeds up the process.

What is the significance of all this for building a sailboard? The process of hardening begins as soon as the catalyst is added to the resin and accelerator: great care should be taken in measuring the amount of catalyst exactly as prescribed, and in mixing it in thoroughly. As a rule it involves 1–2%, sometimes 3%, of the quantity of resin; exact proportions are stated in the manufacturer's instructions. Where thickener or fillers are used, note that mixing percentages refer to the amount of *resin* used, not the total mixture. The accelerator is usually supplied premixed in the resin in the interests of safety, since it reacts violently if in direct contact with catalyst.

The hardening time for polyester resin varies according to the amount of hardener (catalyst). The normal 'pot life' is around 25–30 minutes at a room temperature of 20°C (68°F). Pot life is the period from when the catalyst is added to the resin until the beginning of the hardening process, after which it must not be disturbed or worked. This is the point when the resin begins to gel and is no longer liquid. Immediately afterwards it begins to harden, which takes another half to one hour.

It is also possible to add a retarder to increase the pot life, which gives more working time. This is preferable to adding less catalyst since it should never go below 1·5%. (Most fillers also have a slightly inhibiting effect on the resin cure rate.) Retarders saturate the polyester resin at the beginning of polymerization, and the resin then only becomes unworkable when it is at the final stage of solidification.

After hardening, the laminating resin at first remains sticky (tacky) on the surface open to the air. The styrene solvent present in polyester resins evaporates very quickly and after this process further hardening of the surface is no longer possible. But this is not as bad as it sounds, for it makes it possible to continue working on the previously laminated surfaces without preparing or treating them further. This means one can layup on top of them, even after a period of days. A perfect chemical and mechanical bond results between the old and the newly laid up layers. However, it also means that the final surfaces requires an additional step to become hard and impervious, and for this a special sanding coat or flow-coat resin is used.

A drawback of polyester that can be serious for the amateur builder is its strong smell. This is unavoidable, and in commercial workshops styrene levels have to be controlled. If you have to work in your house, you will find it unpleasant, and neighbours may complain if you use your garage.

Epoxy resins

A lot of what has already been said about polyester resin also applies to epoxies. This resin is used in board building with a polystyrene foam core, because this plastic does not tolerate the styrenes or polystyrenes given off by polyester resins. Epoxy can also be used on polyurethane foam blanks.

How does epoxy differ from polyester resins? It has a longer working time or pot life than polyester, and its hardening times are accordingly slower. The resin takes longer to become viscous, gels more slowly, and because of that allows more time to work with the material. The glass fabric can be smoothed straight, adjusted and the resin worked in thoroughly. Epoxy resin also does not shrink. It is stronger than polyester in terms of tensile strength, resilience and fatigue strength. It is also more resistant to the destructive effects of water. Epoxy is the better adhesive and should always be used when fixing straps and inserts into the board.

Naturally, where there is light there is shadow: epoxy is in comparison to polyester resin a good deal more expensive.

With polyester it is sometimes possible to add catalyst (hardener) without measuring it absolutely exactly, whereas working with epoxy one has to be especially accurate. The manufacturer's instructions must be followed exactly and the proportions adhered to. However, this only takes care and is not really difficult. Use really good kitchen, lab or commercial scales so that the ingredients are as exact as is at all possible; purpose-made dispensers are also available, and often more suitable.

Polyester is preferred by some professional builders because they are used to its shorter working time, can arrange ventilated workshops, and because it is easier to polish to a smooth finish. Epoxy is more difficult to sand down and polish, and needs harder abrasives than the usual DIY ones. However, 3M (Scotch) and others do make suitable grits, and car polishing compounds with wool buffers are effective in the final stages. The job will take time, but conversely the extra hardness that makes the resin slower to polish also helps to protect its appearance in use.

No resins are waterproof in absolute terms, and both kinds are commonly given a final sealing coat of two-part polyurethane varnish, though this is not strictly necessary. A board is more likely to absorb water through loose fittings or screws, minute cracks from impact, deep scratches or other neglected damage.

Hardeners for epoxy resin can be allophatic or aromatic chemicals. Some industrial epoxies use the latter, which have a greater risk of causing dermatitis or allergic reactions but may be attractive because of their lower cost. It is important to use hardener of the aromatic type, which is more suitable for amateur use being safer to handle and breathe.

Storage of resins

All resins should be stored in closed rooms at a constant cool temperature. Moisture and dust should not be able to get to the glassfibre or into resin cans, and resin is very susceptible to changes in temperature. Recommended storage temperature is about 15°C (60°F) or less. Most resins come under the regulations for inflammable liquids. The average shelf (storage) life is at least six months, but depends on temperature and whether it is pre-accelerated, as it usually is for amateur use.

Both polyester and epoxy resins can become dangerous if large amounts of resin mixes or residues are kept in a bucket. They generate heat as a result of the chemical reactions taking place—it would not be impossible for them to burst into flame or even explode. This also applies to thick layups, which have been built up continuously (unlikely on sailboards), or piled-up incompletely hardened scrap. In the case of epoxy resin the hardener is extremely dangerous: the stronger it smells the more toxic it is. Inhalation of these fumes, or of the styrene given off by polyester, can cause headaches and illness. It is therefore advisable to have very good ventilation in the working area.

Storing hardeners

Hardeners (catalysts) contain peroxide, which is extremely reactive and corrosive. Storage and dispensing must be in polyethene or glass containers, not in metal. Only the amount absolutely necessary at any particular stage of the job should be in the working area. Larger amounts should be stored elsewhere. Spillage may ignite whatever it lands on, or the rag or paper used to wipe it up: regard such waste as inflammable and even liable to burst into flame. It can be placed in cold water outside the shop. **Hardeners and accelerators should be kept apart at all costs: on no account should they come into contact with each other because of the danger of explosion.**

Health and safety

The working area should always be kept clean. Splashed or spilled catalyst or resin needs to be cleaned up immediately and very carefully. Naturally the workshop should be well ventilated, yet held at the specified temperature range for laminating. There are regulations for the maximum amount of styrenes in the air: these should not be exceeded, and the legal levels may in fact be optimistic.

The materials store should be separate from the working area. Safety goggles and long gloves are essential for handling catalyst, resin and glass. In the case of any chemicals splashing onto the skin immediate action is required, above all if some goes into an eye. The treatment is to flush the affected area, especially eyes, with copious amounts of plain water for 15 minutes. Catalysts with peroxide are especially harmful: again flush the skin or eye with running water, and do *not* apply any ointments. In the case of any splashes in the eye it is advisable to consult an eye specialist, *after* carrying out this immediate first aid treatment. Resin removing cream should be used on the skin, not solvents. Barrier cream applied before working is essential to safety and comfort and helps in washing afterwards.

Understandably enough, smoking

is prohibited in or around the working and storage areas. If chemical fumes cause irritation of eyes or nasal passages, drowsiness or other symptoms, which can lead to unconsciousness, remove the victim to fresh air immediately and ventilate the room. Sanding, filing or grinding hardened resin or laminate throws off tiny particles of glass and resin which are dangerous to eyes and lungs: always wear tight-fitting goggles and a respirator or face mask that filters the air you breathe.

A very few people have skin or other reactions to laminating materials of various kinds, and these may not appear on first use. To minimize the chance of their occurring it is essential to eliminate as much direct contact as possible—usually by skin contact, breathing fumes or dust, eating or drinking while working or before washing thoroughly, or using solvents to clean the skin.

Tools and Workshop

Polyurethane foam can be worked on with an ordinary saw, a band-saw, Surform planes and files. A hot-wire cutter is only suitable for poly-styrene blanks, and instructions for making this device are given in this chapter. If this is too much trouble polystyrene can also be cut and shaped with a hand saw or band-saw, Surform, power planer or by hand sanding. A keyhole, coping or sabre saw and a sharp Stanley knife are useful for cutting stringers, etc.

The following tools are necessary for cutting the foam blank, shaping it, designing, laminating the outside, and for finishing.

—Electric orbital or belt sander
—Hand saws for wood, wide and narrow bladed
—Surform planes or open-cut files (long, short and curved)
—Sanding blocks (20, 50 and 100 cm); production paper (3M)
—Sandpaper of assorted grades up to very fine wet-and-dry
—Long ruler or measuring tape
—Long flexible edge (curtain rail or thin wooden batten)
—Spirit level and set-square
—Small pocket/block plane for the stringers
—A jig or adjustable fitting to prevent the electric drill or router going too deep
—Circular grinder or router to cut out the skeg wells
—Stanley knife and extra blades
—Padded blocks or workbench, to lay the board on
—Breathing mask to cover nose and mouth, safety goggles, barrier cream for hands and forearms
—Protective gloves, rubber or disposable plastic

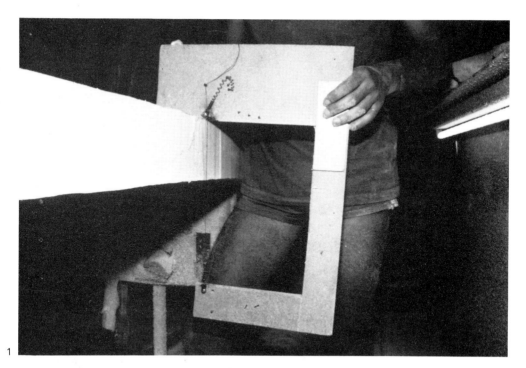

A hot-wire saw made from scrap chipboard, with a spring to tension the wire (courtesy S P Systems)

Making a hot-wire saw

—Coloured felt pens, pencils
—Air temperature thermometer
—Stippling (coarse) brushes, mixing pots, acetone or epoxy solvent
—Wide and narrow masking tape
—Rubber squeegee, sharp scissors, plastic sheeting, paper, rags
—Polishing buffer and compound, wide paintbrush

Many tools will already be found in the home, others can perhaps be borrowed or hired, and with a bit of skill one can make the rest, for example the stand for working on the board. This is not difficult, and a width of 60 cm (24 in) is normally quite sufficient.

If the board is being made with polystyrene then it really is recommended to make a simple hot-wire saw. The EPS foam blocks are made and often sold in large dimensions and consequently a lot of material must be cut away. This is far easier and more exact with a hot-wire saw than with a normal woodworking saw. (Polyurethane foam is cut only by sawing, with a bandsaw or hand saw.)

To reduce 220V or 110V to 40V a transformer is necessary. The cutting wire should be chrome-nickel 0·3–0·6 mm (nichrome), or even better (but more expensive) wolfram. One also needs three pieces of wood, an

elastic shockcord as a tensioner, screws, an on-off switch, and an electrical connector so the saw can be separated from the power lead.

The temperature of the cutting wire is dependent on its length: the longer it is the colder, or the shorter it is the hotter. If the budget allows, buy a dimmer with a 0–200V range to connect in between the trans-former and the electrical supply socket. With this it is possible to regulate the power supply to the transformer and thus the heat of the wire.

Before even starting to cut out the core for the board you should do some test cutting and practise the technique. The tension and tem-perature of the wire must be set up

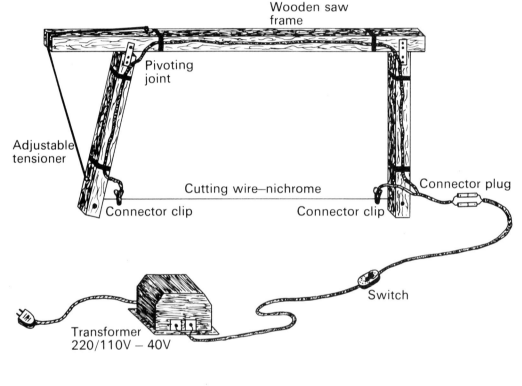

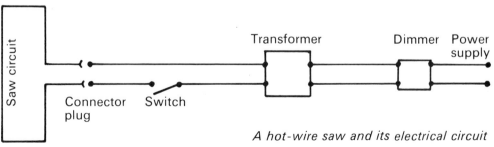

A hot-wire saw and its electrical circuit

2

Working in the cutting and shaping area. A great deal of dust is produced and it must all be cleaned off the board and surroundings before starting on the graphics and laminating. Note the very well padded, stable stands for holding the board on edge or flat, and the low-level strip lighting. (courtesy S P Systems)

first. Do not fix the wire too tightly as it can become overstretched on heating and may break during cutting. The tensioner compensates for thermal expansion and contraction. The wire must be guided and kept moving throughout the cut, otherwise it will melt a groove in the foam. Heat-proof guides or templates can be set up on either side to keep the cut square and give the desired curves. A helper to steady the blank and cutter can be very useful.

The working area

The area of the workshop should be at least 15 square metres. Obviously in a cool climate it must be possible to heat the room, and in any case to maintain its temperature at a constant level and still ventilate it well. Because of the inflammable materials, chemical fumes and later on paint spray, heating by means of any type of gas burner with a flame, or by electric fires, is highly dangerous.

Positioning the lighting is a science in itself—it takes a bit of trouble but is well worth it. It is important to be able to spot the smallest unevenness, to disclose rough edges or check the overall outline. You need perfect lighting especially for shaping the core. The ideal arrangement has proved to be a row of 40W fluorescent lights suspended over the middle of the board. Other lamps should be mounted on both sides along the workbench. If they are adjustable so that they can be raised and lowered and switched on independently, then the conditions are perfect for precision work: unevennesses will be revealed immediately by light shadows. With the overhead lighting, the edges and the general outline can be checked. With the side lights a trained eye can assess the evenness and the curve of the bottom or deck shape.

It goes without saying that the tools should be so arranged that the correct ones are at hand for each working procedure. Searching is very unnerving, especially when it is important to work quickly. Tools hung on the wall in a special order save a lot of time.

Some working procedures are very sensitive to dust. If the work-room is large enough it should be divided in two, cutting off part of it with PVC sheeting to separate the laminating work from the heavy dust-producing procedures. The separate room for storing hardener, resin and paint should also be lined with PVC sheet, but it should not be so tightly sealed that fumes or damp accumulate. The glassfibre storage also has to be free of dust and any dampness.

It must always be possible to remove spilt resin, if only because of the danger of fire. If an ordinary table is used as a workbench then it is advisable to cover it: resin can then spill over without permanent damage. The sheeting need not be of a heavy grade, and can be cheap enough to discard and burn after the job is done.

It has already been mentioned that the chemicals involved can be disagreeable and dangerous, particularly catalyst but also resin, especially when it has been mixed with hardener (catalyst) and before the reaction has entirely finished, and also the cleaners and styrene solvent. A cold water tap and sink for flushing off the skin or clothing, and a bucket of water for any rags or paper that have been used to wipe up wet resin, are also desirable, in the interests of safety.

Spilled resin can be absorbed in earth or sand. Catalyst must not be mopped up but be diluted with large quantities of water, as it can react with organic materials, such as rags, paper, plastics, wood or your eyes and skin.

In case of fire, dry-powder extinguishers can be used on resins, release agents, accelerator, polystyrene and polyurethane foam, mould cleaner, acetone and solvents. Polyurethane foam will give off highly toxic fumes in burning.

Do not attempt to use water as an extinguisher once fire has started, *except for catalyst which must be extinguished with water.*

Shape Design

The diagrams on the following pages showing different board shapes and some working drawings should enable you to plan, draw and build your own board. Your ideas may fit in with the professional builders' ideas and experience; however, some questions must first be answered. What do I want from the board? How well can I sail? Where will I use this board? Am I only a light winds sailor, or must there be a strong wind and waves to have any fun? From these questions, and your experience of trying others' boards, it will be possible to identify the shape most suitable for you. The volume of the board is related to the user's body weight and should by no means be overlooked.

The lighter a board the quicker it will race through the water; it is correspondingly fragile and susceptible to damage, yet sensitive. Its inertia makes a board easier to sail, as well as width and a stable shape. If in doubt, aim for a less radical and more durable concept. It will keep its value longer, and probably give you just as much fun.

Drawing the plans

Having decided which shape suits your skill and sailing technique, it should first be drawn to scale on millimetre graph paper. This step is very important because stencils or

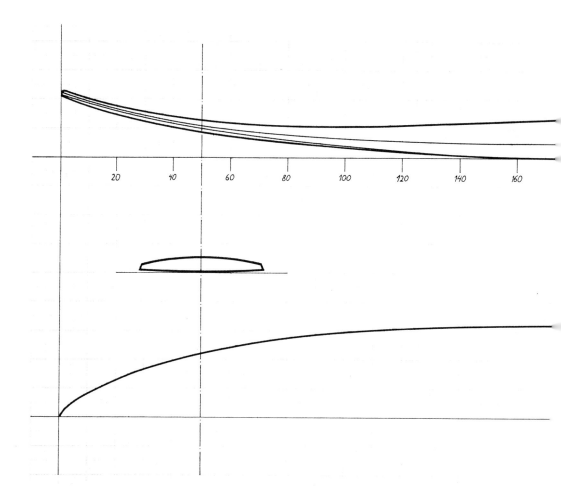

templates are produced based on these drawings with which the board is shaped from the raw foam block.

First divide up the millimetre paper by marking off 'stations' every 2 cm (representing 20 cm) as shown here. Pencil in the length of the board and the height of the 'scoop' of the upswept bow and also how far back along the bottom the scoop should run. Carefully join these points with the flexible rule, making sure that the curve of this bottom line is fair —

smooth and without bumps or kinks. Now decide how the volume of the board is to be distributed along its length and mark these points on the drawing. From the millimetre divisions on the paper it will now be possible to see exactly how thick or thin the board will be at these points. When they are joined the deck line will emerge, which completes the profile seen from the side. To round the deck and bottom, take the vertical depth measurements at a number of stations

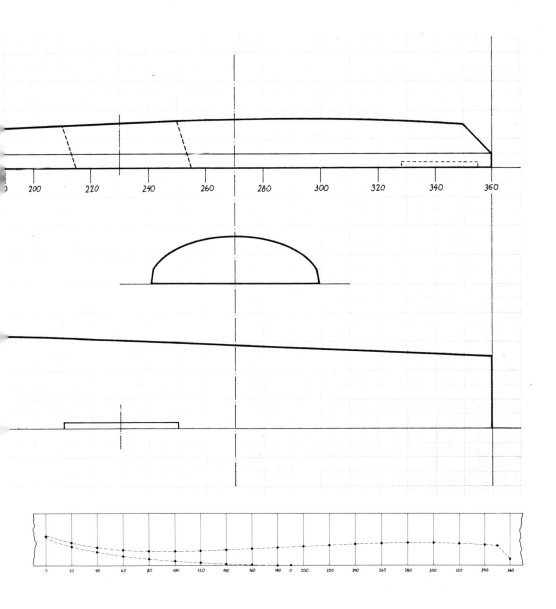

and use them to draw cross-sections, at the same time considering the width (beam). The points of maximum beam also have to be drawn and faired to develop the outline in plan view, giving further borderlines for shaping the foam blank.

Making a stencil of the edge outline

After the plan view drawing on millimetre paper is finished, transfer the design in the ratio 1 mm = 1 cm onto cardboard or strong tracing paper, taking the exact measurements of the width from the centreline for each station on the graph paper. Then stick the stencil onto the foam block with tape so that it cannot possibly slip, and draw the markings onto it. After having marked the points join them together with your long flexible rule (a thin wooden batten or plastic curtain rail), thus outlining the edges of the board.

Making a stringer template

A stencil or template for a stringer is made in the same way. The vertical height of the deck line and the bottom line are measured from the baseline at each station and transferred in the same 1 mm = 1 cm ratio onto the stencil paper. The points are joined and faired with the flexible rule, then the stencil is cut out along this line with scissors and stuck carefully and exactly onto the unprepared stringer wood. The outline is traced or sprayed with paint and the stencil removed. The stringer must be cut out along the sprayed border, but about 1–2 mm wider, with a fine narrow bladed saw.

Templates for cutting the blank from a larger block of foam are held to the sides of the block and used to guide the hot-wire saw. So they have to hold their shape, resist the heat of the wire, and have smooth edges. As the pair of cutting templates must leave the blank never smaller than its maximum designed thickness, their dimensions are taken from the maximum thickness measurements, i.e. on the centreline, with a slight addition for the amount to be lost by later smoothing down.

Stringers, which will be glued lengthwise in the core to strengthen it and define its shape, are cut to match the designed curve and thickness at their particular place in the board. For example, if there are three stringers, the middle one would be taken from the vertical measurements of the bottom and deck lines along the centreline, as is shown on the side-view drawings and as for cutting templates. However, the other two stringers are offset from the centreline and have to be taken from different dimensions, i.e. those at the corresponding distance out from the centre on the drawings. These can be taken from the cross-section drawings and obviously the more of these the better, especially if the contours are at all complex.

Preparation and Stringers

The amount of preparatory work that will be necessary before you can begin shaping and laminating depends on the type of foam blank you have. Slabs of various sizes, pre-cut to approximately the desired dimensions or even already stringered and partially shaped, can be obtained from kit suppliers such as Shore Boardsailing. Some custom builders will provide professionally shaped cores to your special requirements, ready for home completion. At the other extreme, you may prefer, or have to, do everything yourself. Foam obtained from a manufacturer or building supplier is likely to come in a size which needs cutting down. Buying jointly with other people will save money and materials as the large blocks can make more than one blank.

Stringers are made from sheets of wood veneer or plywood about 4–6 mm thick. Epoxy is preferable to polyester resin for bonding the stringers and sections together as it is a much better adhesive and stronger. It is more work to build in stringers, but worthwhile in terms of durability, less flexing and a better fit for the mast and daggerboard fixings.

Tools

— Long straightedge
— Bandsaw (try a joiner or carpenter), otherwise a hand saw
— Hot-wire saw, only for polystrene
— Face mask
— 4 mm wood veneer
— Epoxy resin, perhaps colouring to add to the resin
— Yardstick or sliding gauge, long spirit level, measuring tape.

Working time: two hours, with a twelve hour waiting time.

3

Cutting out the blank

Cutting the edge outline from a flat blank, resting the frame of the hot-wire saw on the top surface as a guide to keep the cut vertical. The essential centreline has been drawn on the blank. Tricky shapes like these wings require practice on scrap foam before they can be cut as smoothly as this. (courtesy S P Systems)

Once you have decided on the maximum dimensions of your board, the stock foam can be cut down to a workable size. This must allow extra material as insurance against damage or early mistakes. It is easy to take off a little more foam, but once you have removed too much the problem is more serious. So the first step is to check and recheck the drawings, and then the dimensions you have taken off them, ensuring particularly that the side view, plan or profile view, and sections through the board at various points all match exactly.

Either measure off and mark your raw block or slab for sawing, or make up hardboard templates as guides for hot-wiring. Cutting guides and marks should allow for the material that will be lost in the cut, regardless of method; they should also be slightly over-size in all directions to keep that safety margin mentioned above. Templates need smooth edges for a smooth cut, and a few holes drilled through for long nails to pin them firmly against the surface.

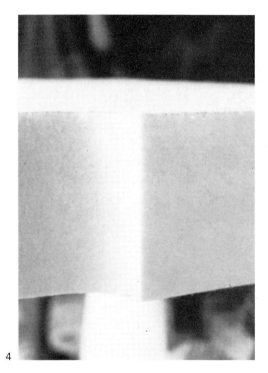

4

Stringers

Wood veneer 4 mm thick, often red cedar or mahogany, serves best for stringers (the thinner the lighter) and five or more stringers may be built in as desired. Two is the normal number, and quite sufficient. The stringers are cut out with a keyhole or coping saw or a sharp knife from a pattern taken from the drawings. The grain runs lengthwise. They will be smaller than the blank, but slightly larger than the final shape to allow later adjustments.

Avoid placing them just where skegs will be, and if there is a daggerboard use two stringers so they will not be weakened by cutting through.

Slice the blank lengthwise, very evenly, where they will be inserted. Cover the prepared wood on both sides with a good layer of thickened epoxy, such as SP115 with glass bubbles, and also the foam beyond

Hot-wire cutting of extruded polystyrene or the bead type is much easier with two people, especially on a large piece. Set it up so it is firmly supported, can't shift or wobble, and with enough room for you and the saw to keep moving steadily to the end of the cut. Most important, practise first on some long scrap pieces.

The alternative is sawing, with a fairly fine-toothed hand or power saw that won't tear out chunks. Again, test first: too fast a saw speed will melt the plastic. A timber supplier or joiner may be able to cut thick pieces for you on a table or band saw. Polyurethane *must* be cold-sawed because of the toxic fumes it gives off when melted or burned.

5

6

pieces are clamped together: a few locating pegs or pins through the veneers into the foam may help hold them in position. Confirm also that both stringers are an equal distance in from the ends of the blank. The clamps are as yet only loosely tightened.

Check the symmetry of the top of the board with a spirit level; it could have warped or twisted. If necessary adjust the individual pieces until they are symmetrically positioned.

Only now may the clamps be screwed tight. It is necessary to check the individual lines and surfaces of the hull once more with a yardstick, measuring tape and spirit level. It may be that the bow of the blank is too thin so that the stringers peep out, in which case the blank can be fitted to the stringers by bending the foam upwards, if you work quickly and the resin has not begun to set. Because the foam will tend to spring back, and to achieve a good result overall, it is important to leave the clamps long enough on the board. Leave them overnight before they are loosened; the room should have a temperature of 20 °C (68°F) at least, and during curing it should have been maintained at the specified temperature.

It is possible to colour the stringers, using pigments and colourings which can be mixed into epoxy or polyester resins. This gives a fine coloured line along the stringer which becomes visible in shaping, and helps with accuracy as well as being decorative.

the ends. Masking tape (for epoxy), or straps tightly round the board with its edges protected against crushing, serve for clamps. If you must use polyester resin, work quickly as it sets faster.

The level of the stringers in the blank must be the same, i.e. they have to be laid the same constant depth in from the surface. To ensure this measure the distance from the outer surface of the hull down to the stringer, all the way along the board from stern to bow, with a short measuring rule or sliding gauge. Obviously the veneers should not come out through the deck surface, and should not move when the

Shaping the Blank

Shaping is one of the most important procedures in building a sailboard. It demands more precision than any other: you should take your time and work slowly and accurately. A lot of time can thus be saved later on as the work that follows will be kept to a minimum.

Tools

—Power planer and/or orbital sander
—Long Surform planer or hand plane (at least 1 m long)
—Small block plane or spokeshave
—Hot-wire saw, hand sanding block
—Hand saw or compass saw with a narrower blade: fine-toothed
—Sandpaper (120 grain), emery cloth, foam rubber
—Carpenter's long spirit-level
—Flexible long edge such as curtain rail or plastic strip
—Set-square
—Measuring tape, felt pen, spray varnish.
—Masking tape, prepared stencil(s), cardboard for smaller section templates
Working time: two to four hours.

Shaping

Polyurethane foam can have a very hard outer surface resulting from high pressure during manufacture. This must be removed, and it can be done more easily with a planer than

7

8

9

stringers to be 3—4 mm under the surface. It is important that the surface remains even and doesn't become dug out or lopsided. The hull must remain symmetrical: check frequently with the long spirit-level (Fig. 8).

The last few millimetres to the stringers are taken off by hand with the help of a long sanding block or Surform (Fig. 9). The basic principle here is—check, check and check again! At this point it must be ascertained that the board edges and surfaces run parallel at both ends and that its surfaces are completely even (Fig. 10).

The small plane is needed should the wood stringers have to be planed down to the level of the foam. The foam is softer than the wood veneer, and care must be taken that it remains undamaged while planing. Plane with the grain of the wood and avoid tearing out splinters. When both the stringers and the foam are flush the centreline can be drawn in once more, still working with the board bottom-up. Measure the

with a belt or orbital sander— working with the latter is very tiresome (Fig. 7). Turn the board bottom-side up and plane the whole surface carefully with a hand plane. Only go down far enough for the

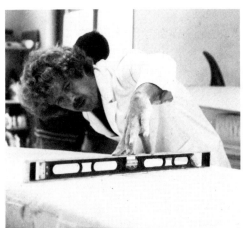

10

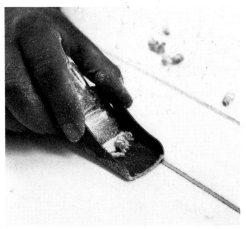

11

12

distance between the stringers at several points and establish the middle of the board using a felt pen, measuring tape, and the long straight-edge (Fig. 12). This centreline is important as it serves continually as a baseline for the work. With it one knows where to lay the stencil and where the skeg boxes, mast step and daggerboard well are to be positioned.

Tape the stencil (made from the plan view drawing) onto the centre-

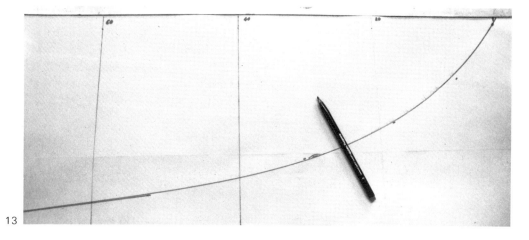

13

14

15

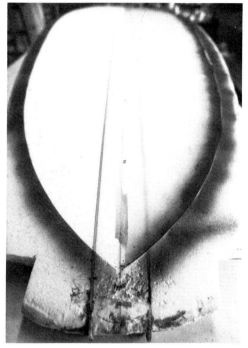

line of the blank. In this way it is now possible to mark the plan view outline directly onto the foam core (Figs. 13–14).

How can this be done most easily? Usually one draws a felt pen line following the half-stencil, but there is an easier method: spray along the edge of the stencil with a tin of spray varnish. Part of the varnish wets the blank and part of it colours the paper of the stencil (Fig. 15). Watch out that the varnish doesn't eat into the foam: this can happen very easily with polystyrene as spray paints contain solvent which attacks it. Therefore try them out first on a piece of scrap foam, and only spray on polyurethane. Otherwise – or if in any doubt – use a coloured felt

marker pen.

Turn the half-stencil over. The centreline again serves as the symmetrical axis and with the stencil on the other side it is a simple matter to mark in the other half of the outline (Fig. 16).

After removing the stencil the outline is sawn around with a hand saw. It is always advisable to add a 2–3 mm margin to the drawn edge. Because the line of the deck edge cannot be seen, the saw is held somewhat slanted so that the blank is wider towards the surface away from you, which will become the board's deck (Fig. 19). At all costs avoid sawing into the deck.

The outline can be straightened with the help of a long sanding block

18

or a hand plane (Fig. 20). There should be a few millimetres to play with, and after the planing or sanding work towards the drawn line. Continually check the right-angle which the edge or side should have to the hull (Fig. 21).

19

20

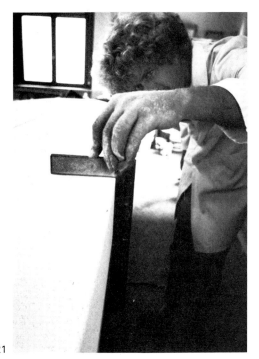

When this is done, take a sanding block (which should be at least 1 m long) and work along the outline in long smooth strokes. The aim is to get a smoothly running fair curve (Fig. 22). Take plenty of time: it is important to be very meticulous here.

It can be seen from the plans that the bottom of the board should be rounded and swept up in the bow. In order to sculpt this shape out of the flat blank, first of all draw a guideline on the edge surface of the blank (Fig. 23). This line is again taken from the scaled-up drawings. A flexible wood batten is suitable for this or a curtain rail or plastic strip serves just as well (Fig. 24).

Stand sideways and work from the centreline out towards the lines drawn on the edge. Use the long hand

21

22

plane or power planer (Fig. 25). Do one side first, bevelling it down for final sanding. Then plane and sand the opposite edge. The line on the side is your reference for width as well, so don't sand it off.

If you haven't got a good eye, then work on the second side with the help of small templates—the board should have identical contours on each side! Make small cardboard templates every 10–20 cm by scrib-

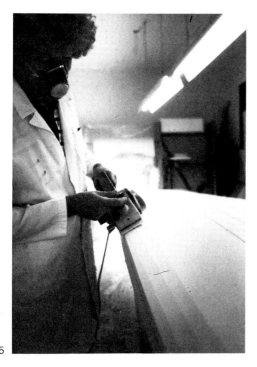

25

ing them with a scribing tool and pencil and check the second side with them. It is well worth going to the trouble even if it takes more time, for it is impossible to correct the underlying shape later on. It takes a lot of work to correct mistakes even if only small bumps and un-evennesses occur. To check, sight along the board from the ends and across it. Deviations from straight lines, or any wobbly curves, can be discovered, and adjusting the side lighting will help.

Finally take a longer sanding block and 120 grain sandpaper: smooth the surface further with this. A piece of sandpaper or sanding cloth is better for the curves. Move it gently with both hands over the edges to make them evenly rounded (Figs. 26, 27).

Shaping a complex bottom with steps or concaves will be more difficult. With a single concave across the board the stringer(s) may have to be planed down. Double channels and just one stringer (or a flat area between them) are easier,

Bevelling the edge along lines marked a uniform distance in from the edge (above). This gives a curve ready for the flexible sanding block and final smoothing with fine production paper (below).

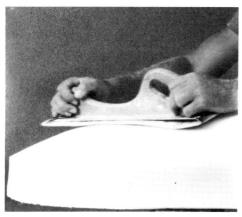

26

27

but making both sides identical will take care. First mark the edges and centrelines of the concaves and stay within the lines, using a curved Surform and then sandpaper with a foam rubber backing pad. A length of cardboard tube a few inches in diameter, with the paper held round it, also works. Pros make shaped sanding blocks. Check the depth by measuring down from the spirit level set across the board, and be very sure not to take off too much foam.

Steps also need precise marking out and cutting, using a Stanley knife or perhaps a router with guide. Setting up a jig or guide strips to control the depth and prevent wobbles is worth the trouble. One problem when laminating over steps (and to a lesser extent concaves) is that the glass will tend to spring up out of the corners and dips, leaving voids and rounding off edges. This can be overcome, but take a good look at professionally made boards before you design anything that will be impossible to lay-up properly and still keep its intended shape.

Shaping the deck is a similar process. Turn the board over so that its deck is now upwards. Mark the centreline, lateral lines for width and rocker, and edge shaping guidelines. The contour can be drawn exactly onto the stern. A planed-down deck ready for smoothing with sanding block and then strips is shown in Fig. 29.

It is difficult, because some flair and skill are necessary, to sand the

28

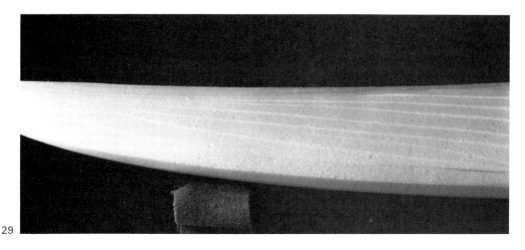

29

chosen stern shape and curves perfectly even. To help in this very fine sanding one can make a simple tool: laying the sandpaper on a piece of foam rubber gives a better surface. Take care also when working on the deck shape because the foam is easier to sand down than the wood veneer of the stringers! The stringers must constantly be approximated to the level of the foam: for this use a small block plane. Again plane with the wood grain not against it so that the foam does not get damaged or the stringer roughened or torn. However, very little planing-down should be needed, if the drawings, stencils and fairing were carefully done.

Any voids that appear in the foam can be patched by fitting in a piece of foam with resin, and sanding it down neatly.

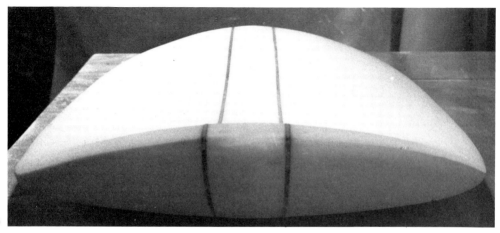

30

Colour and Graphics

One of the signs of a custom board is its colourful graphics, and this is where you can give your imagination free rein. However, there are still a few constraints to keep in mind. Many spray paints, including those for cars, contain solvents which will attack polystyrene foam and polyester resin. Unless your board has a polyurethane core, for painting or air-brushing directly on the foam water-based acrylic paints from art supply shops will be necessary. Some felt-tip pens are also safe. (Always test paints on scrap first.) Otherwise you can wait and apply them after one of the epoxy/glass layers has set overnight and been lightly sanded (180 wet-and-dry), which will protect the polystyrene. The glass goes colourless when wetted-out and the design shows through. For sharpness, some people paint on the final lamination, which gives only the thin gelcoat as protection against damage.

Tools

— Scotch masking tape, newspaper
— Paints: watercolours or spray
— Brush, spray cans or spray gun, felt pens
— Fixing varnish or lacquer

Some other possibilities are Letraset symbols and lines, or using gelcoat colour paste in the fibreglass layup. (Keep some for later repairs.) Moulding shops have a range of such products. Japanese rice paper, or the fine glass tissue sold for fibreglassing, can itself be coloured, protected with fixative lacquer such as ICI Clearcoat and incorporated in the layup. This might be the best idea if your decorative skill is uncertain: any mistakes or unsatisfactory results on the cheap tissue can be scrapped and you can try again. Correcting work done directly on the foam or glass is much harder, and

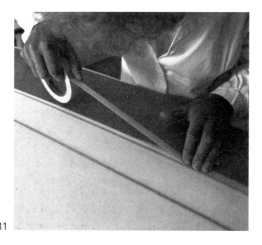

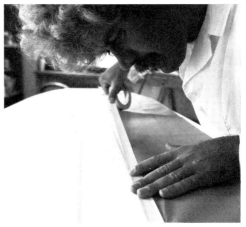

31

32

may involve removing the material. Consider also some form of identification to deter theft. Striped or geometrical patterns that are striking but don't require freehand work can look complicated but with careful masking are not hard to build up.

The first step is to get out *all* the dust from the earlier work, and brush the board clean. Dust will give a poor result when painting and laminating.

All surfaces which are not to be coloured immediately, except for the stringers, are covered with paper and masking tape. Spray mist or dust will otherwise stick to everything. Colour shadowing looks very attractive and decorative. It is not difficult to let the colours blend into each other using spray paint. It is sprayed on thickly from one end then allowed to fade out slowly. In the 'faded' area the second colour is carefully and lightly sprayed over the first one and then more densely towards the other

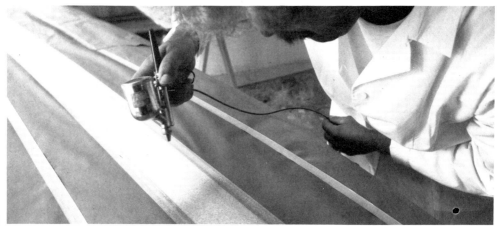

33

end. In the middle area the particles of the second colour sprayed over the first blend into a very harmonious colour shadowing.

Perhaps you know a graphic artist or draughtsman from whom you could borrow a spray gun with a small compressor. The very fine paint spray permits subtle and fascinating designs and very clever shading. Only you must know how to use it and it's rather expensive to buy. Certainly the spray-on paints in cans are simpler, very effective, not expensive, and easy to use.

When felt pens and spray paints are used it is very important to fix the colour after drying, although it is always good to protect the whole board with fixing varnish once the design is finished. Otherwise lines and contours could easily run while applying the resins. Again, when polystyrene is being used the fixing varnishes could contain solvents which would damage the material. Test the reaction first on a piece of leftover foam painted with the colours.

Ultraviolet (UV) light in sunlight affects colours. It is preferable to use lighter shades since dark colours are warmed more by the sun. The heat under the laminate then becomes stronger and any non-slip wax applied to the deck goes soapy and slippery. Blistering can also occur.

Apart from this, you would be aware that in the course of time and under the influence of UV rays, boards become slightly yellow. Thus blue colours, for example, become light green and the reaction of the other colours varies; some dark colours may absorb more UV and be more affected by it. Epoxy resin tends to yellow, though some systems such as SP115/215 have better clarity and light stability.

Finally, any carbon fibre reinforcement must be glassed over with a light, opaque resin with filler. The black carbon becomes very hot under the sun's rays.

Laminating

The characteristics of resins were discussed in the chapter on materials, in particular the properties of polyester and epoxy resins, one of which will be used for the GRP outer skin. The shaped foam board, fragile and unprotected, must be made unbreakable with a laminated surface. The method of laminating is the same for epoxy and polyester resins, the only difference being in the accuracy and speed which are demanded. In either case, however, concentration and working speed are required, otherwise the resin could begin to harden on the half-finished board and make it useless.

As already mentioned, polyester resin has a very short hardening time (pot life). Using epoxy resin, there is more time to remove bubbles and make clean edges. The disadvantages are that it takes a laminate with this resin longer to harden completely (which doesn't often matter much), and the ratio of resin to hardener must be very exactly controlled.

Materials

— Glassfibre cloth, possibly carbon fibre reinforcement or Kevlar cloth
— Scotch masking tape, Stanley knife, extra blades
— Small coarse-haired brushes, rubber squeegee, roller
— Rubber or plastic gloves, goggles, breathing mask, barrier cream
— Resin, hardener; acetone or xylene; epoxy solvent
— Weighing scales or liquid measuring dispensers

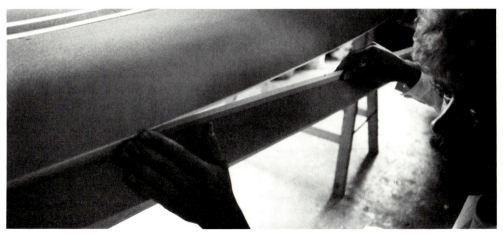

—Surform or open face file, sanding blocks
—Mixing containers and stirrers, scissors

Working time: about an hour if working continuously from one layer to the next. Allow longer.

Laminating technique

The underwater surface is laid up first. Cover the other side of the board edge with 1 in. wide masking tape (e.g. Scotch) suitable for resin work. (A chemical reaction can take place with polyester: the adhesive on some tapes dissolves and forms a compound with the resin which then prevents it from hardening and the adhesive sticks to the blank.)

The tape must be stuck on 3 cm back from the edge, beyond where the bottom and side (rail) edge meets the deck. The aim is to overlap the glassfibre layers when laying up the deck (Fig. 34). In other words, there is a double thickness of glass along the edges of the board, which is desirable because that area is both under special stress and susceptible to damage. If the masking tape lies exactly under the line, and another layer of wider tape is stuck on top of it, the deck will in addition be protected from dripping resin; plastic sheet taped on to further protect it is a good idea. The edge of the masking tape closer to the GRP layup defines the cut-off line of the overlapped glassfibre, so it should be placed neatly and precisely.

Preparing the glassfibre
Now lay the glassfibre cloth over the bottom surface and smooth it out by hand (Fig. 35). Trim any superfluous material exactly along the edge of the hull with scissors, overlapping the tape a little. Lay the second layer of glassfibre on top, smooth it by hand or with a wide brush, remove the creases and trim off the excess precisely, this time leaving about 5 cm extra, but no more than that. Otherwise it is difficult to fit the cloth around the edges and corners and make it stick. Fitting is easier if

the cloth is cut in a little on sharp curves and points, in V-shaped darts, depending of course on its weave and weight. Once it has been soaked with resin it is harder to cut with scissors, and wastes time. The corners of the stern and the bow tip can be strengthened with a patch of glass. Mark the centreline on the cloth ends, then roll it up.

Mixing the resin Have all tools at hand before mixing the resin with hardener. You need: paintbrushes, rubber squeegee, spatula or fur roller (not so good), gloves, goggles, mask, scales/dispensers, Stanley knife, and a second person to help if available. You should already have the barrier cream on your hands and forearms and also gloves as protection against glass splinters, while cutting the cloth.

The amount of resin needed depends on the size of board and your skill in wetting-out all the glass and removing the excess. A rough guide is about 1 litre per side per layer. So, for a 3 m board with three layers of glass cloth on the deck and two on the bottom, as described here, about 5 L will do, plus a little more for fittings. However, breakages have led many amateurs to use four layers per side, especially for wave boards. Extra resin in the lay-up doesn't mean it's stiffer, only heavier and more expensive.

Laying-up Before mixing the resin and hardener together, ensure that they have reached 20°C (68°F), and that the blank is dry and at the same

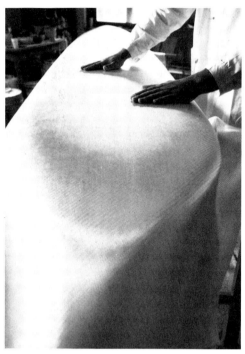

35

temperature. Paint the mixed-up resin on the foam, quickly but without leaving any dry areas. Then lay on the glass and pour on more of the resin, not all at once or it will run off and be wasted. Brush and squeegee it evenly over and into the glass, but with care not to pull it around. A foam roller can also be used, especially on the edges. It saves time if a helper stipples resin into the edges with a coarse brush (Figs 37, 38).

As soon as the glass is well soaked with resin, work the latter into the mats with the squeegee. All air bubbles should be worked out and the glass saturated evenly all over. This has been achieved when the separate fibres of the glass are no

36

longer visible; all light (dry) patches should disappear. Having got this far scrape or 'wipe' the superfluous resin out of the fabric, working from the centreline outwards (Fig. 39). This surplus resin can be put back into the bucket. Check now whether the glass cloth on the edges and corners is still lying correctly on the core with an even overhang all round. Where there are air bubbles stipple a little more resin onto the

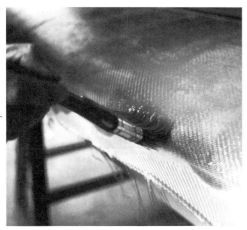

37

38

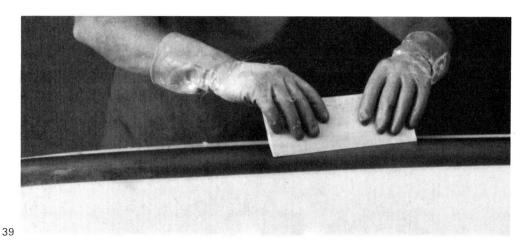

39

spot and remove the surplus once more with a spatula or paintbrush. Leave the laminate to harden.

Depending on the temperature, but at the latest after 1 to 2 hours, the rough edge of the laminate can be cut at the masking tape (Fig. 41).

This is best done with a Stanley knife or sharp chisel when the GRP is at the 'green' or 'toffee' stage of partial hardening. If the laminate has been made with epoxy resin one must wait two to three hours. Hard, that is completely set, laminate can only be

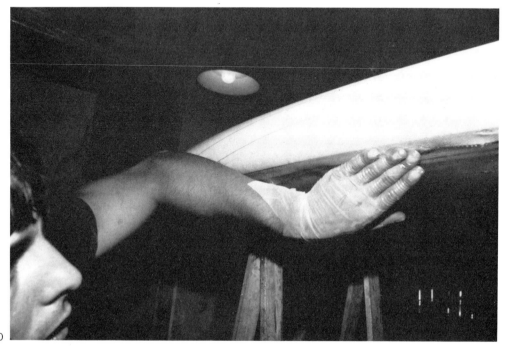

40

cut with some difficulty so do not wait too long. Where the rails are sharp, you may want to cut along the corner. Smooth the rough edge left with a Surform or sanding block (42) before putting on the next layer, which overlaps it slightly.

Most people laminate the bottom completely first, applying the next layer of glass right after the previous one goes non-sticky (green) but before it is really hard. This is much faster, but it means that the over-lapping layers of the deck and bottom don't interleave at the edges, which would be the strongest lay-up. Turning the board over after each side was given one layer would mean waiting at least overnight for it to harden properly, to prevent dent-ing or distortion. Then the laminate would have to be prepared for the next layer by degreasing with the prescribed solvent, drying off and sanding. This is not popular (though it has to be done if you need to lay-up in stages), so edge strength is achieved by overlapping successive layers and sometimes running uni-

directional glass along the top and bottom edges inside the last layer of cloth. Where extra strength is want-ed, usually around the mast foot, daggerboard slot, skegs, and on the thinner tail, glass tape or sometimes Kevlar is applied before the last overall layer.

Laying up the top of the board is identical to doing the bottom. Ad-hesive tape is stuck on 1 in. below the edge line so that the laminate will overlap the layer already applied (Fig. 43). Glassfibre cloth is again

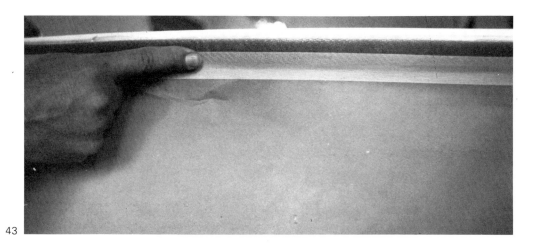

43

laid on the board and smoothed flat. With a slight difference: three mats are laid on the deck area where the sailor stands, the last one sometimes of heavier cloth (say 8 oz), as the board must withstand more pressure in these areas.

Tools should be cleaned immediately after laminating and before resin has started setting. Less flammable solvents such as zylene or methyl alcohol are better than acetone. Only stripper (Nitromors) will remove resin that has set.

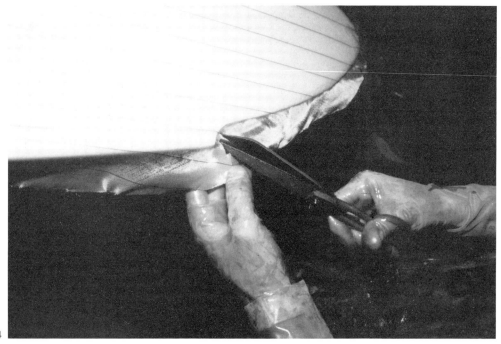

44

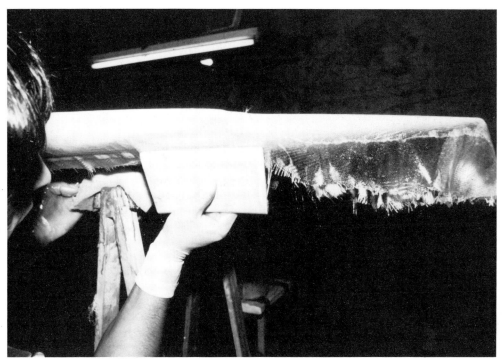

Concave shapes and sharp corners like these wings are complicated to laminate well and cuts in from the edge of the glass allow it to fit the curve. When working the resin through the glass (below) too much pressure will pull the cloth out of the curve, leaving a void. This is one of the outer layers, which overlaps the others for strength and a good final finish after sanding. The high stands used here provide a comfortable working position with a good view of the top and bottom at the same time. (courtesy S P Systems)

Kevlar reinforcement

The golden, apparently pliable Kevlar cloth is actually quite difficult to work with. It is hard to cut, quickly dulling scissors or knives, so it is even more necessary than with glassfibre to achieve a perfect fit before resin is applied. Furthermore the cut edges fray, and once hardened the laminate cannot be ground, filed or sanded smooth as it frays up and roughens. Thus even more careful laying-up is needed than for glass, and the cut edges are best taped down while it hardens. Where glassfibre cloth is laid up in one or two layers over Kevlar it provides an outer surface that can be ground smooth, but otherwise the rough area where the top and bottom layers of Kevlar overlap on the rails can be covered with glassfibre tape and resin, which can be sanded normally.

Completing the Board

There is only one thing to be really careful about when putting in the skeg boxes, mast step, centreboard well, towing eye and safety hooks: there ought not to be too much resin in the recesses at any one time. During hardening, so much heat can be generated in the narrow cavity that is sawn or machined out that the foam around it can melt or burn. An internal void then forms which is not visible from the outside, and only makes its presence felt during the first hard sail, when the loosely anchored fitting could be torn out.

Careful cutting so that slots and holes are a good fit, and a long curing time, are essential. Choose fittings with a large surface in contact with the core and skin, to spread stress and increase bonding area. You may be able to extend non-ideal ones with screwed-on wood strips, even so they reach the opposite skin. Plastics do not bond very well, especially polythene; fibreglass (GRP) and polycarbonate are better. First degrease them with solvent, abrade and score all over to get a 'key' and clean again; then set them in with thickened epoxy.

Tools

- Rule and pencil, set-square
- Router or grinder
- File or very sharp knife, such as Stanley knife
- Glassfibre cloth strips, epoxy resin, scissors, spatula
- Sandpaper, rubbing compound, solvent cleaner
- Glass balloons; colloidal silica
- Rags, masking tape, protective paper or plastic sheet
- Drill, hacksaw, sabre saw
- Protective gloves and goggles, breathing mask when sanding

46

47

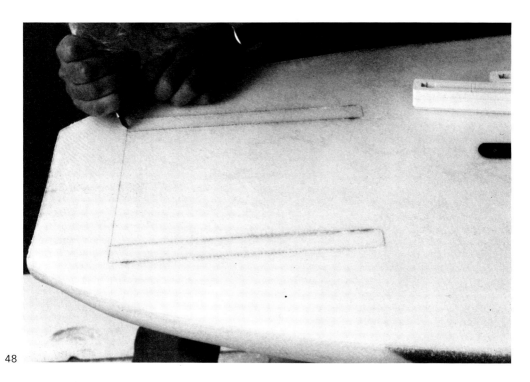

Skeg boxes

With a rule, the centreline on both sides of the board is again found and pencilled in. Obviously skeg boxes or wells only project from the bottom surface. Their position is more or less established, but the number is your decision. Skegs must be positioned exactly in relation to the centreline (Figs. 46, 47), even if they are thrusters. This is also most important for all the other fittings.

The laminated skin is cut out carefully with a Stanley knife and can then be removed easily (Figs. 48, 49). The shape and depth of the skeg box is already known and a slot to these dimensions must be cut into the foam. The easiest method is to use a router, or a grinder in an electric drill, but it is also possible with a knife or narrow sabre saw, though more trouble (Figs. 50, 51).

The depth of the skeg wells must be doubled on jumping boards or those which are to have long skegs, and the boxes themselves must be strong. It is especially important for these wells to be very stable laterally otherwise they will loosen quickly and could actually come out of the board. Also, loose wells will work and open up, letting water into the board. Screw, glue or laminate a piece of foam or wood to the skeg well so that the depth is doubled. Some board builders add an extra layer of glass or Kevlar reinforcement to the stern on designs where the fins will be highly stressed.

49

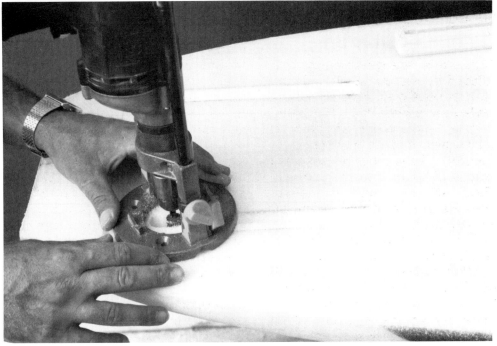

50

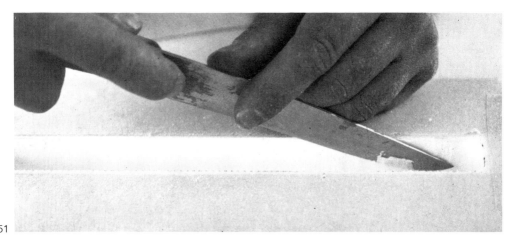

51

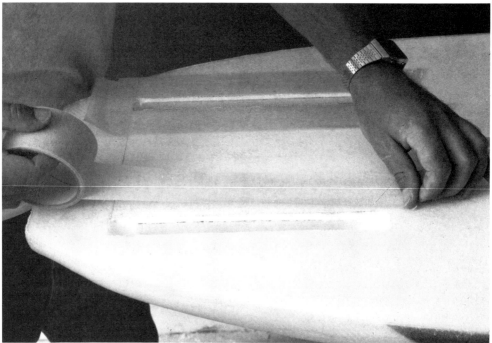

52

The hollow is now clean and accurately prepared; stick plastic or masking tape on the area around the hole at a distance of 4 mm (Fig. 52). This is necessary because the 4 mm boundary must now be worked to a bevel of under 45° with a file or

53

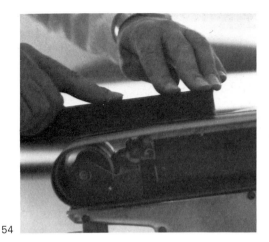

54

sharp knife (Fig. 53). This will be the basis of additional lateral support for the skeg box.

Before continuing it is important to ensure that the skeg wells and the surrounding surface are free of grease; to be absolutely certain use sandpaper and cleaning solvent, and sand and score the fittings all over to get a key (Fig. 54). As soon as you have covered the slit in the skeg box with tape and cut two glassfibre cloth strips to the size of the cavity (Figs. 55, 56) the main part of the

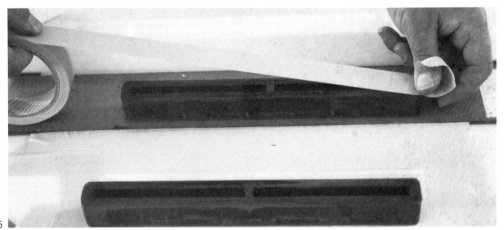

55

56

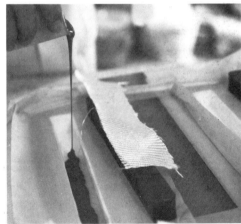

57

procedure can be started; it is similar for bonding in all the fittings on the board.

Mix 100 g of resin (for deeper wells somewhat more) and thicken well with microballoons or aerosil. If you want to give the 45° bevel a visual effect colour can be added (Fig. 57). The hollow for the well is half filled only with the resin mix, the strips of glassfibre laid on top, and the part pressed in with great care (Figs. 58–60). Surplus resin pushes up and out of the well and must be

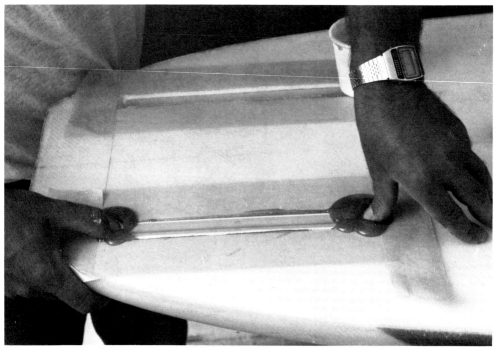

wiped off immediately from the board. At this stage it can be seen how accurately the slot was cut with the router or grinder. The skeg box must fit tightly and squarely into the board. If a knife has been used a quick check is necessary. Remove the tape from the slit in the skeg well, put the skeg in carefully, and check with a set-square whether the position of the skeg is right. If not, then correct it. Hold the skeg well in position by running tape across between the board edges and over the skeg tips when they are exactly vertical.

It is worth trying the skeg box for fit before starting to glass it in: if the cavity has turned out to be too big then proceed one step at a time and take care not to pour too much resin into it at once. Pour in a little at first,

press the skeg box in and support it up exactly straight, and wait until this resin has hardened. Then pour a little more into the gap. The heat of the reaction must be kept to a minimum because it quickly melts away the foam, so the volume of wet resin at any time has to be kept very small.

Mast step

The mast support is, after the skegs, under the most strain as it has to withstand a lot of pressure and sideways thrust. Choose mast tracks or sockets that sink deeply into the foam; it is even better if they can be bonded to the laminated skin on the bottom. To spread the forces on the

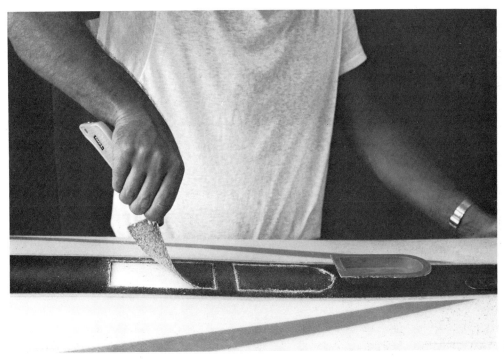

61

62

working loose. Or the sockets can be supported by cup-shaped fittings set into recesses in the foam with resin paste (Figs. 61–5).

If sockets fitting in cups are to be used, draw their positions on the deck and drill or rout out their holes (Figs. 65–8), which must match the width of the fittings. Do not go too deeply, and whatever happens not through the board! These parts must also be absolutely grease-free. If the sockets are open at the bottom they can be closed with a little tape. The recesses are again surrounded by tape so that the board is not smeared with resin (Fig. 63). Thickened epoxy is used to bond in the cups (Figs. 64–5), which must be left to cure thoroughly before drilling.

mast wells, they can be inserted in holes drilled in a hardwood block which is glassed into the board on the centreline, thus giving a larger bonding area as well as less chance of local crushing of the foam core or

The next steps are much as before.

63

64

65

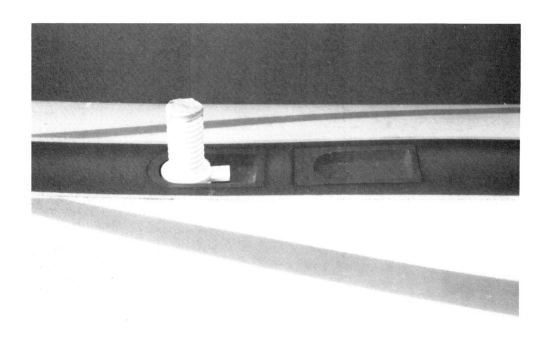

66

68

69

Small strips of glass cloth are cut and wrapped around the socket fittings. The thickened resin is put into the holes drilled through the cups or mast step, the mast sockets with the glassfibre round them are slowly turned into the board and the surplus resin is removed (Figs. 64–69). By this method the mast well can be guaranteed completely waterproof. Fill the well up to the top with cold water, which has the effect of retarding the reaction of the resin and protecting against possible heat damage.

It has often happened that the fittings have literally shot into the air; if there is too much resin in the recesses too much heat develops during hardening, and it burns away the foam around the fitting well.

Centreboard well

Even though centreboards (pivoting) or daggerboards (non-pivoting) in funboards or wave boards are now looked down on, experience has proved otherwise. If the centreboard is adjustable sailing comfort is considerably increased, and it can be very useful to have enough windward performance to get where you want and sail home. If wells with completely retractable boards are not available, anyone capable of building their own sailboard won't be put off by having to build a centreboard well and sketches with guidelines are in a later chapter.

A neat one-piece fitting can be made by laminating in fibreglass over a male mould, as in Fig. 70. (See

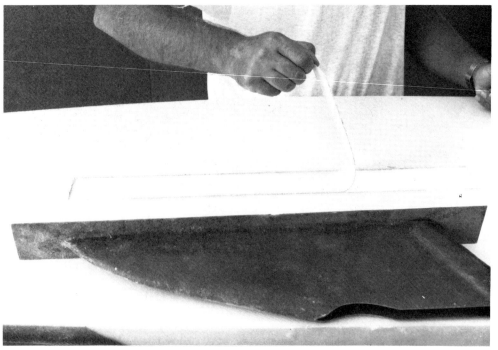

70

71

72

65

also Figs. 96–101). Commercial fittings are in two parts with flanges on both sides. Though this makes them more rigid and gives additional fixing area, the plastics used do not always adhere as well as GRP to the resin.

Draw in the position of the well on the bottom and deck. The section to be cut out for this is marked out on all sides and the GRP skin removed. Cutting down the centre with a hand saw helps orientation and in cutting out the slot with a keyhole or sabre saw (Fig. 71). It again pays to cut exactly 'as it makes a tighter fit and saves work later on.

Tape and masking paper or plastic are stuck down 1 cm back from the sawn-out gap in the deck. As on the skeg boxes, this gap is bevelled

to 45° with a file or knife after the slot has been finished, making a wider bevel than for the skegs. The constructed well can be set in as soon as it is dried and sealed with waterproof varnish or polyester or epoxy resin. Glass cloth must be slipped in sideways between the well and the foam—this works well using a spatula (Fig. 72). Once more the same precaution with the resin: the bevel may not be filled all at once with the thickened and coloured mix (Fig. 73). This should be in two or three steps, otherwise the resin's exothermic heat would melt the foam and possibly run through.

After the resin has hardened on the deck side of the board, the bottom of the well can be glassed in. Check with a set-square against something stuck through the well whether it is parallel to the centreline and exactly vertical.

Safety line and towing eye

A fitting for a safety line in the bow is only necessary when the board is meant for sailing in breakers, jumping or extreme conditions. A hook or eye is built into the bow and the rig and board are connected there by an elastic shockcord. Should the rig separate from the board it will then function as a floating sea anchor, pulling the bow into the waves since the board on the surface drifts faster than the rig. This keeps the board from blowing away from the sailor who is now in the water.

For more ordinary use, it is always prudent to have an eye near the foot of the mast to take a rig retaining line (mast leash), as it can come out during a capsize.

In some countries a towing eye is compulsory on commercially made boards, though the examples seen are often too small to be much use. A bow eye is worth building into any board, not only for towing but also to help in lashing the rig to the board, carrying, and lashing it down on a car-top or outdoor rack. A standard dinghy fitting may be usable, if it can be bonded successfully; or a hole shaped in a block of wood that is then glassed into the bow would be very strong and protect the GRP from chafe. Make the hole big enough to take easily ropes of the large sizes used on yachts (for towing) or several lashings.

Defects and Remedies

If the skeg boxes and skegs, or the mast socket with the rig, land on top of your head the first time out, it is nearly always obvious why. The work was not quite exact and careful enough. Too much resin gathered in the hollowed-out recesses for the skeg boxes or the mastrack, because either the cavity was too large (too much resin) or the resin was poured in too quickly. The resulting high temperature burned or melted the foam. The same fault may develop along the centreboard well, sooner or later, as it is laterally stressed. It may take awhile to become evident and therefore is something to watch for.

Such loosening of components is not a total writeoff by any means. It is possible to repair by foaming-out the cavity again with two-part poly-urethane foam. This also means, of course, that the whole procedure of glassing-in the mast wells or skeg boxes has to be started again from the beginning. This time one knows exactly what is important! Corrections are very seldom necessary with accurate work.

Air bubbles which have got caught in the laminate are simply aesthetically unpleasant. They are critical only if numerous, or if small patches of GRP laminate come off. Even so, a small trick can help here. Using a disposable plastic syringe, resin can be injected afterwards into the defective area, in between the laminate and the foam core; although if there is liquid in the bubble it is better to cut away the skin down to the foam, dry it out and patch the hole with new glass and resin.

Finishing the Surface

Your board has at last taken shape, but you should not overlook the finish on the surface. This is a very important step, after building in all the fittings, to give maximum speed, a uniform and professional appearance, and protection against weathering and the gradual accumulation of dirt, and to prevent water absorption.

Sanding and polishing take time and patience, and you should allow at least 48 hours after each coating for it to cure hard. Maintaining dry warm conditions (18°C minimum) is essential throughout, and the board has to be kept dust-free. Mask off all slots and fittings, and protect the side you're not working on with tape and plastic sheet. Footstraps may be fitted (see next chapter) before final varnishing.

Materials and tools

— Wide brush, squeegee, sanding blocks, wet-and-dry sandpapers 200–600
— Orbital sander
— Salt, sand or commercial non-slip finish e.g. Peel Ply
— Breathing mask, goggles, gloves, barrier cream
— Acetone or ethylacetate, or epoxy solvent
— Rubbing compound, rags, fine polish (T-Cut), wool buffer
— Wide paintbrush (not losing hairs)

For polyester:
— Wax in styrene
— Finishing coat resin or air-drying
— Two-part varnish (optional)

For epoxy:
— Epoxy finishing coat
— Two-part polyurethane varnish (optional)

Finishing techniques

A board with a polyester skin is slightly simpler to finish as its foam core will not be styrene soluble, and the finishing or sanding coat is really an extension of the laminating. It can be painted over with polyester resin which has been mixed with 2% wax in styrene. The wax floats on top of the brushed-out resin and prevents the styrene in it from evaporating too quickly, thus enabling the resin on the surface to harden throughout and preventing tackiness. The same effect can be achieved by laying a thin plastic or cellophane film on the freshly coated surface. However, since this must be laid quickly and skin-tight, ripples and creases can easily form, resulting in the long run in unnecessary additional sanding which could penetrate the protective finish that is being sought.

As soon as this surface coat is completely hard and cured the board is sanded and cleaned with fine-grain sandpaper. As with all resin and varnishing work, it is important to keep the surface free of grease from handprints. A final thin, low-viscosity coating of 20–25% styrene added to surface coat with wax, or of whatever type is recommended by the resin manufacturers, is then painted on with a soft non-hairing brush (Fig. 78). NB: Do not brush over any area twice.

Surfaces that will not be non-slip can then be sanded smooth and

74

varnished with two-part poly-urethane, such as SP2000 which gives UV protection and waterproof-ing. All varnishes leave only a thin layer after drying, so apply three coats and use only very fine polish-ing compounds afterwards.

Boards laminated with epoxy need a different treatment, and harder fine-grade abrasives (try 3M industrial products). Smooth off all rough places on the last glass laminate, wet-and-dry sand all over to a smooth matt surface, and clean with solvent. The resin and hardener for the final 'gloss' coat is probably the same system used for laminating (e.g. SP115/215), perhaps thick-ened with colloidal silica for easier spreading.

After hardening, wet-sand with fine grades (400, 600) but not through the resin. Then polish with car rubbing compound (e.g. T-Cut).

Three coats of PU varnish (SP2000) protect against UV light and water and can be fine-polished with a wool buffer.

A non-slip surface can be made by sprinkling salt onto the last resin coat while it is still wet. A more uniform result is obtained by pressing nylon Peel Ply into the resin, then pulling it off after it has cured.

The easiest alternative is to apply a non-slip paint such as Intersurf.

A further possibility is to utilize the texture of the glassfibre or Kevlar cloth itself. Finish coat (without filler) is spread over the deck with a squeegee and all surplus removed by pressing hard. While the surface is thus sealed, the indentations in the weave of the glassfibre remain and are naturally antislip without adding anything except wax (Figs. 79, 80).

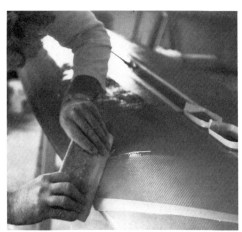

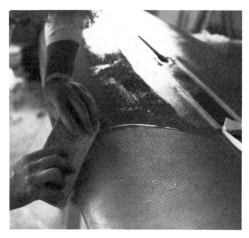

79

80

Footstraps

Mounting footstraps is done after the surface is finished, otherwise there would not be a perfectly flat deck surface, which is necessary for smoothing out resin and removing the excess with a squeegee. Footstraps are only necessary for jumping and speed sailing. They give more control and a more secure feeling since the sailor becomes one with his board; his slightest movement is transmitted directly to it.

However, there is a certain element of risk with footstraps. If they are too big you can pull muscles or ligaments, sprain joints or even break bones in falling. It is far better to squeeze into tightish straps than to have too much leeway for the feet. When the straps have been made for shoes and you sail barefoot, your feet will slide in too far or come out accidentally. Therefore decide in advance whether the footstraps are to be for shoes or bare feet.

Materials

Method 1: plastic plugs, sliding gauge, drill, screws.

Method 2: sabre saw, small spatula.

Method 3: tape, covering paper or plastic sheet, spatula, glass cloth.

Footstraps of non-stretch webbing, with padding. Masking tape, epoxy resin, glass balloons as thickener.

Working time: one and a half hours.

Mounting the footstraps

Having been decided on footstraps, the question is how many and where to position them. It is absolutely non-sensical to plaster a board with footstraps! Twelve or fourteen are as superfluous as twelve loudspeakers in an ordinary room. Three or at the most four suffice, if well placed. You must have sailed your board in order to establish the positions; it is important to know where you stand, which depends on your technique and the wind speed. Take the time to be absolutely certain of your customary positions and where you actually need straps.

Three methods of mounting are described here. The simplest and quickest is with plastic plugs (like those used for screws) glued into the board with resin. If the straps turn around the screws, use a fluted metal lock-washer between the plug and the strap. In any case, screw through a plate on top of the strap ends.

Plugs Measure the diameter of the plugs with a sliding gauge and drill a matching hole in the skin and foam core. Take care to drill only to the same depth. The plugs are then glued in with thickened resin (Figs. 81–5).

Laminating into the board The surest, neatest and most durable method has proved to be glassing the footstraps into the board. This demands very precise work to get a

81

82

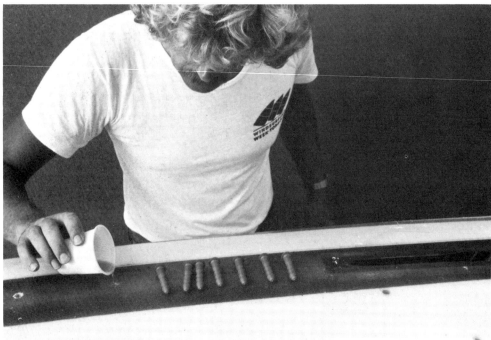

83

84

85

tight fit and ensure against melting the foam. With the small jigsaw cut narrow slits 4 cm deep, put in resin and insert the straps. Burn holes through the ends for the resin to penetrate and grip.

Laminating onto the board The best looking method by far is fibre-glassing the straps onto the board, although it is work-intensive (Figs. 86–95). The exact area to be

88

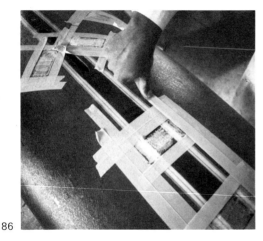

86

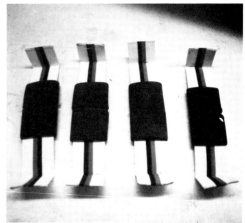

89

87

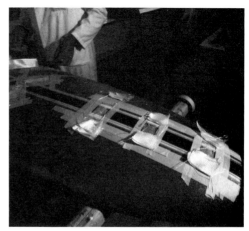

9

91

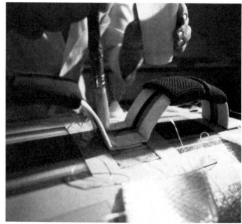

92

93

laminated is marked out and masked with tape and all the surrounding surface and the part of the straps that will be exposed is carefully covered.

With a small grinding bit, take off the smooth surface for a key (Figs. 86–7). The area should extend well beyond and around the strap ends. Avoid trapping air, which can happen when glassing-in, by rolling and pressing the patches over the ends of the straps. The layers of glassfibre holding down the straps become successively larger as each is over-lapped by the next, for the best bond.

The usual width for footstraps is around 14 cm (5½ in.) and from this one can calculate the amount of webbing necessary. Obviously it is best to try out different straps on existing boards to get the best fit and comfort, and then allow plenty of material at the ends.

There are footstraps sold which give way under strain—so-called safety straps, similar to hand loops on ski-poles. However, they still need improvement.

Finally, remember that a board without footstraps is always better than one where they are badly positioned.

Carefully wipe off all excess resin and allow at least 48 hours for curing. You can provide the neces-sary warmth over the straps with hot water bottles covered over with newspaper and a blanket, refilling them several times a day. It is important to get the best bonding possible on footstraps—as on all fittings—because they take a lot of

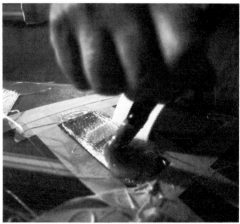

strain that is concentrated on a very small area. For appearance, smooth off rough edges afterwards and paint with resin or polyurethane varnish.

And finally . . .

Your work is finished, but the board is not: the resins have still not entirely finished curing. This process goes on for some time, during which the board develops maximum strength, durability and resistance to damage and scratching. Lay it flat and level so it won't develop a twist, and leave it another two weeks in a dry *warm* place, out of the sunlight. Then take your photographs, while it still looks its best.

Making a Daggerboard Case

A wooden male mould is necessary for this method. The casing is laid up in fibreglass over it and then removed after hardening and curing. Chipboard of the type supplied with a thin facing of plastic veneer serves best. The correct thickness (related to the centreboard itself) is the most important criterion when buying, however, and ordinary chipboard or any solid wood can also be used.

With the help of a keyhole saw cut out the outline shape of the inside of the centreboard well and round off the edges. Then bolt it with wing nuts onto a wood base which should be 4 cm wider all round. The extra piece which will make the groove for the board's pivot pin should be attached to the side of the mould in such a way that it remains inside the well on removal from the mould. This is necessary because the

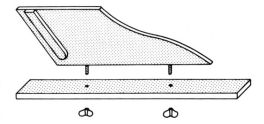

96

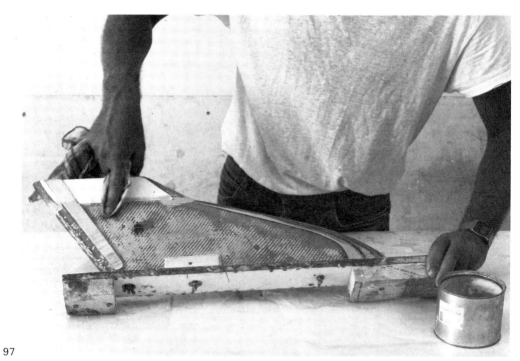

97

98

82

piece making the groove must be separated from the bigger wooden mould in order to withdraw it from the wide end of the finished centreboard casing. The small piece is pulled out from the smaller end of the slot (see diagram).

Before beginning with the laminating the entire mould and base must be sanded smooth and coated with release wax (parting agent) and polished.

Next mix the gelcoat resin and paint the shape thickly with it; gelcoat can be coloured. There is time to cut the glassfibre mat while the gelcoat is hardening. This material is very convenient because it can be easily shaped or stretched around corners and edges. Two layers are enough; the average weight is 200 g/sq m (6 oz).

Before laying up the glassfibre make the polyester resin thixotropic with thickener and brush the sharp edges of the mould really well with it. The aim is to avoid dry pockets between the gelcoat and the laminate next to it.

When this preparation work has been finished the glassfibre mat can be laid on and saturated with resin. In practice it's best to lay up first one side and let it set and then the other. Doing this avoids the danger of the glassfibre on one side slipping or being pushed away while working on the other side and forming bubbles because it wasn't hard enough. At the 'toffee' or 'green' stage of hardening the surplus edge of the laminate is trimmed off around the mould's base with a sharp knife or chisel and the base removed. Thus

100

101

a 2 cm wide flange remains on the centreboard well moulding. This ensures the necessary overlap around the case when it is built into the sailboard (Fig. 101). At the same time trim the rough edges on the smaller end, and make sure they're not stuck together, closing the slot.

It is now possible to separate the wood mould from the well: run a thin knife carefully round between the laminate and the mould. If the two do not separate then dip it in warm water: the mould release wax becomes liquid and it shouldn't be any trouble to loosen the well from the mould.

Fig. 102 shows a wooden male mould for laminating a mast cup, the support for a mast socket fitting. Laying-up is done in the same way as for a centreboard well, using this shaped wooden block. The glass edge is trimmed neatly at the 'toffee' stage of curing.

Learn how the Experts do it...

STANFORD MARITIME

Member Company of the George Philip Group
12 Long Acre, London WC2E 9LP, U.K.

Beginning...

FASTER BETTER WINDSURFING
Preuss, Taaks, Winbeck

The latest methods for faster learning, using correct techniques from the start so you improve faster. Result: a sound basis for funboard and advanced sailing. Based on the highly successful teaching system of the German VDWS (Association of Windsurfing Schools). Also covers equipment, racing, rights-of-way with other craft, tides, rescue, emergency repairs. Paperback, 122 pages

WINDSURFING TECHNIQUE
Stickl and Garff

A unique approach, using sequence photos. Starting from the basic stance and manoeuvres, with detailed analysis and intruction to take you up to advanced sailing. This is the book to improve your performance.
Hardback, 180 pages.

Racing...

WINDSURFING RACE TACTICS
Noel Swanson

All the board-against-board situations that occur in racing are analysed here. Fully illustrated, and shows the best courses and moves in response to opponents. Tactical problems of starts, upwind legs, mark rounding, reaching and downwind, and finishing are all covered. Also broader strategy, tide and currents, and increasing your fitness.
Hardback, 157 pages.

WINDSURFING RACING TECHNIQUE
Philip Pudenz and Karl Messmer

Two of the world's top sailors analyse and demonstrate fast tacking and gybing, mark rounding, 720° turns, trapezing, starts, tactics and courses.
Hardback, 180 pages.

Waves and Freestyle...

HEAVY WEATHER WINDSURFING – ON FUNBOARDS AND SINKERS
Jürgen Hönscheid and Ken Winner

Handling strong winds, waves and breakers, is explained by two of the best-known windsurfers. Covers fast turns, wave-riding, speed, jumping, racing in breakers, equipment, and the winning techniques for every type of course. Extensively illustrated.
Hardback, 120 pages

101 FREESTYLE WINDSURFING TRICKS
Sigi Hofmann

Starting with preparatory exercises and some easy stunts, this book makes it possible to work up to the more daring and spectacular ones, and includes advice on linking them into a programme. Stunts are graded according to difficulty and wind strength and arranged so you build on techniques already learned. Illustrated throughout.
Paperback, 120 pages.